ROYAL
OBSERVATORY
GREENWICH

T0136545

The Sun

Brendan Owens

Royal Observatory Greenwich
Illuminates

First published in 2021 by Royal Museums Greenwich,
Park Row, Greenwich, London, SE10 9NF

ISBN: 978-1-906367-86-2

At the heart of the UNESCO World Heritage Site of
Maritime Greenwich are the four world-class attractions
of Royal Museums Greenwich – the National Maritime
Museum, the Royal Observatory, the Queen's House and
Cutty Sark.

rmg.co.uk

A CIP catalogue record for this book is available from
the British Library.

Typesetting by ePub KNOWHOW
Cover design by Ocky Murray
Diagrams by Dave Saunders
Printed and bound by CPI Group (UK) Ltd, Croydon,
CR0 4YY

About the Author

Brendan Owens is a passionate science communicator, with a keen interest in and knowledge of astronomy and physics. Currently Astronomer Emeritus for Royal Observatory Greenwich and Open Science Coordinator at Science Gallery at Trinity College Dublin, Brendan prides himself on weaving fascinating stories to share his passion for science with school and public audiences alike.

Entrance to the Royal Observatory, Greenwich, about 1860.

About Royal Observatory Greenwich

The historic Royal Observatory has stood atop Greenwich Hill since 1675 and documents over 800 years of astronomical observation and timekeeping. It is truly the home of space and time, with the world-famous Greenwich Meridian Line, awe-inspiring astronomy and the Peter Harrison Planetarium. The Royal Observatory is the perfect place to explore the Universe with the help of our very own team of astronomers. Find out more about the site, book a planetarium show, or join one of our workshops or courses online at rmg.co.uk.

Contents

Introduction

Can you name a bright star? Maybe you'll think of the North Star, Polaris, that is often mistaken as the brightest star in the night sky, given its special name. Or, if you have a deeper knowledge of astronomy, you might say Sirius, which is in fact the brightest star in the night sky. However, we often neglect to think about the Sun being the brightest star in the daytime sky. It's not particularly big as stars go, but its proximity to Earth is unrivalled. Also, its appearance, effects and dominance in the daytime make it feel radically removed from what we automatically think of when we hear the word 'star'.

You might not consider it important to know the inner workings of this star on our doorstep, but civilisations have relied on a good knowledge of the Sun's changing position in the sky for various aspects of our daily life, for example for food production, from hunter-gatherers to we modern-day citizens. On a more instinctual level we shiver when a bank of clouds obscures the Sun, and we feel its warmth when we emerge from shadow into the light. There's no doubt that the Sun is fundamentally important to our lives. This was not lost on ancient civilisations, where this book begins, and over the millennia record-keeping of the Sun's activities has evolved into scientific investigations. In more recent centuries, scientists and engineers have looked to apply the outcomes of these investigations to generate power to replace fossil fuels (solar panels,

nuclear fusion research). In much of the modern world today we may as individuals think less of seasonal food production and more about the news feeds on our devices, but here again our star, the Sun, is key. Solar eruptions jettison matter in our direction and can interfere with satellite communications and power supplies.

The following short read explains how humanity has interpreted the Sun and its power over millennia – from a sacred and mysterious orb of light to a cosmic engine that we rely on to survive and thrive. It's an intriguing tale that also provides a greater explanation of how stars work.

Sacred Sun

... just as the battle was growing warm, day was all of a sudden changed into night. This event had been foretold by Thales, the Milesian, who forewarned the Ionians of it, fixing for it the very year in which it actually took place. The Medes and Lydians, when they observed the change, ceased fighting, and were alike anxious to have terms of peace agreed on.

Herodotus, 5th century BCE

Long before the invention of the telescope or the job of 'scientist', peoples across the world recorded the motions, position and appearance of the Sun. In this account from the Greek historian Herodotus, there is no doubt that the armies involved were fearful of what we now understand to be a total solar eclipse, when the Moon temporarily covers the face of the Sun.

Today, standing on the shoulders of giants, we have centuries of scientific knowledge to tap into. A quick search online or chat with a modern-day astronomer can wash away most superstitious thoughts. For ancient civilisations, however, the Sun was wrapped in total mystery, but they knew then, just as we do now: the Sun is life.

If someone could flick an off switch for the Sun, it would take around eight minutes for the last sunlight to complete its 150-million-kilometre journey to Earth. No sunlight means no heat. No sunlight to feed the leaves of vegetation – the vegetation that feeds most life on Earth directly or indirectly. Without the Sun our planet would be just another barren rock in space.

Even without this deeper scientific understanding, the fundamental importance of the Sun to life was recognised. This led to ancient civilisations worshipping the Sun as a life-giver. In turn, all manner of traditions and ceremonies were created to protect it and 'summon' it back. Sun worship might bring to mind thoughts of gruesome human sacrifice in great temples constructed by the Aztecs, but that civilisation belonged to a relatively recent period in human history (1300–1521 CE). Instead, we start our journey of exploration by looking back to the Neolithic period, over 5,000 years ago, where we'll find an example that combines Sun worship with an early understanding of the Sun's motions over the course of a year. Predating both the Great Pyramids of Egypt and Stonehenge in England, Newgrange in Ireland is one of the most ancient architectural examples of predicting and using the Sun's motions for ceremony. The site is a tomb whose full purpose is yet to be understood completely.

Two of the key milestones in Earth's trip around the Sun each year are the summer and winter solstices. These mark the days with the most hours of daylight and least hours of daylight, respectively. At Newgrange, a narrow 'lightbox' was included in the stone structure to let sunlight through at sunrise on dates around the winter solstice in late December every year, illuminating the long inner chamber of this passage tomb.

This might not seem like a big deal when you first think about it. We might consider the Sun simply rising in the east and setting in the west, however, celestial mechanics aren't quite so simple. Depending on where we are on Earth combined with the time of year, the position of sunrise varies significantly from northeast to southeast, and sunset from northwest to southwest.

To realise this change people had to observe the movement of the Sun in the sky daily and look for any changes. Today we know the factors that cause these changes are the tilt of the Earth and the changing distance to the Sun as we orbit around it.

Taking into account these factors, the position of the Sun in our sky changes across the year. If you were to take a photograph of the Sun every day when it is highest in the sky for where you are (local noon), you would end up with the figure-of-eight shape seen in Figure 1. This is called an **analemma** and tells the story of Earth's journey around the Sun.

The smaller of the two loops is when we are furthest from the Sun, which occurs in northern-hemisphere summer. This point of our ride through the Solar System is the slowest so you don't see big changes in the Sun's position happen quickly. The opposite is seen in winter when we're closer and so we speed around the Sun faster, creating a larger

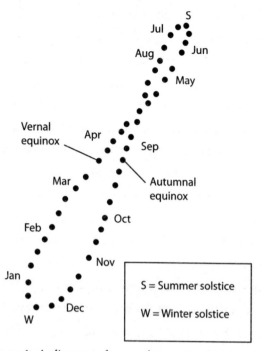

Figure 1: A diagram of an analemma. Each dot in the figure-of-eight pattern shows the Sun's position at different times of the year in the northern hemisphere. The Sun appears higher in the sky during the northern-hemisphere summer, moving lower as we move into winter. The larger loop shows how the Sun's position changes rapidly between measurements. At that time of year the Earth is closer to the Sun and therefore travels faster around it. The opposite is true for the northern-hemisphere summer when the Earth travels further from, and therefore slower, around the Sun. This results in a more compact loop.

loop. The extreme ends of the two loops show the winter and summer solstices. On these days the Sun traces out the shortest and longest arcs across our skies, respectively.

Finally, near the middle of the figure-of-eight shape are the positions of the Sun at the two equinoxes – vernal and autumnal. At these points in the year, we experience roughly equal hours of day and night for the planet – once in March and again in September. These can be seen as the tipping points between the seasons of winter and summer and vice versa.

The analemma is something that can be captured through photography with a lot of patience and consistency. Alternatively, some mathematics, a shadow stick and a watch can be used to achieve the same thing plotted out on graph paper. They're both year-long projects, but an interesting hobby to truly connect with this rock we live on as it hurtles around the Sun at 67,000 miles per hour.

Our modern understanding has been accrued through centuries of learning, but the gaining of that knowledge had to start somewhere. While Newgrange is not the only example, it illustrates the intertwining of cultural tradition and one of the first sparks of scientific study of the Sun through repeated observations, predictions and invention.

Returning to more dramatic myths and legends, major changes in the Sun were seen to spell doom or triumph for a kingdom. None are more ominous than eclipses – both solar and lunar. Total solar eclipses take place when the Moon lines up directly between the Sun and the Earth, meaning the Moon casts a relatively small shadow on our home planet. Anyone in the right part of the world, along the path traced out by the Moon's shadow as the Earth spins, will be plunged into darkness for up to seven and a half minutes. This period of complete darkness is called **totality** and is a very eerie experience.

According to ancient Chinese records, a solar eclipse was the Sun being eaten by a dragon and it was surmised that the beast needed to be defeated. Early on that was by the banging of drums, and later, the firing of ship's cannons. Since the symbol of the Sun was tied to the emperors in China, solar eclipses led to some unusual behaviour on their part. They would temporarily become vegetarian and carry out various other rituals to attempt to 'rescue' the Sun. This perceived connectedness drove the analytical prediction of eclipses from observations of the Sun and the Moon over the years.

Chinese records also give some of the earliest descriptions of the Sun that go beyond its simple appearance as a bright disc. During the Shang dynasty, one record of an eclipse in 1302 BCE proclaimed: 'three flames ate the Sun and there was a big star'. This seems to hint at features that erupt from the surface of the Sun but are usually outshone by its bright disc – we'll hear more about those later. Astronomers have also been able to rewind the modern celestial clock to work out that a good candidate for 'a big star' on that particular day was the planet Mercury, easily seen when most of the Sun is blocked out during a total solar eclipse.

While most people will simply have noticed that day turned to night, those who looked towards the event may have also noted a ghostly glow around the eclipsed Sun. An early account of such an observation comes from Byzantine historian Leo Diaconus in 968 CE:

> ... at the fourth hour of the day ... darkness covered the earth and all the brightest stars shone forth. And it was possible to see the disc of the Sun, dull and unlit, and a dim and feeble glow like a narrow band

shining in a circle around the edge of the disc.

Today we call this feature, the outer atmosphere of the Sun, the **corona**. Until the 1930s and the invention of devices to block out the brightest part of the Sun, a solar eclipse was the only way to see and study this ghostly outer layer of our star. Apart from the 'feeble glow' of the corona, observers could also see bright stars and planets around the sky ... if they could drag their gaze from the main spectacle of the Moon totally eclipsing the Sun.

Aside from these major celestial events, under normal circumstances people were extremely limited in what they could find out about the *nature* of the Sun itself. Great strides in understanding the celestial interplay of the Sun and Moon were made, however, before the invention of the telescope, particularly in the works of medieval Muslim astronomers Ibn Yunus, al-Battani, and al-Biruni. Collectively they observed or gathered reports of 176 years of eclipse events seen from modern-day Egypt, Iran, Iraq, Syria, Turkey, Afghanistan and Turkmenistan.

There were two more sights linked to the Sun that could be observed under special circumstances: one whose link remained unknown for centuries, and the other proving that the Sun wasn't always the perfect unblemished circle of radiance that people directed their worship to.

Scattered through documents across written history are accounts of strange rainbows and glowing lights in the sky. Today we recognise some of these as the Northern and Southern Lights, which we'll come back to later in this book. Suspected accounts feature in the Babylonian Astronomical Diaries, a collection of tablets, some of which date back to 652 BCE. Although some interpretation is required, translations of the tablets have revealed entries about red phenomena at night lasting hours at a time: 'a rainbow whose shine was red' and a 'red glow flared up'. By eliminating other astronomical phenomena such as Moon halos and meteor showers, some researchers are convinced these are the earliest written records of the dancing lights that make up the aurora. Normally they are seen near the North or South Poles, but during periods of high **solar activity** they can be seen much closer to the equator. That would explain these records – the Babylon of old was situated in what is now modern-day Iraq, 33 degrees north of the equator.

As for the other phenomenon that was rarely seen without a telescope, it was recorded during forest fires or volcanic eruptions. As apocalyptic and mysterious as that sounds, the reason was simple. The Sun is too bright to see with the naked eye, but if you throw up a thick enough smokescreen (literally) then the Sun can

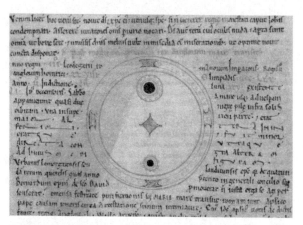

Figure 2: English monk John of Worcester drew these sunspots in December 1128 CE. This is the earliest known drawing of sunspots despite the existence of written records from Chinese observers centuries prior. Judging by the scale of the Sun in his depictions these were particularly large sunspots making their observation easier than other smaller groupings that can occur.

The Chronicle of John of Worcester, 1128–40, Corpus Christi College MS 157, p. 380. *By permission of the President and Fellows of Corpus Christi College, Oxford*

be viewed directly, though somewhat dangerously.

There are accounts of dark blemishes or 'spots' on the Sun as seen during forest fires. Again superstitious rumours ran riot.

What is devouring the Sun now? Is it sick? Is this 'The End'? A monk called John of Worcester sketched what are believed to be the earliest drawings of these **sunspots** in 1128 CE (see Figure 2). The drawings show that he was sketching enormous spots and with very little detail.

A safer technique to view these sunspots was to use an early form of pinhole projection. Today this is still the simplest and safest way to see the Sun indirectly. To make a pinhole projection a pin is used to make a small hole to let sunlight stream into a darkened room. There, light can fall onto a screen and produce a small image of the Sun. This technique was first described by Arabic astronomer Ibn al-Haytham in his *Book of Optics* in 1027 CE.

Before we turn to the advent of the telescope and the means to explore our Sun more deeply, let's have a quick recap. With patience, perseverance, methodical observations, a myriad of interconnected beliefs, eclipses and the occasional fire or eruption, civilisations had observed some special things about the Sun. They saw its changing position in the sky over days, months and years. They witnessed a mysterious glow around its edge during an eclipse, sometimes with 'flame' features. They could, by projection or dangerously through a smokescreen, view large blemishes that would come and go on its disc.

It's easy, therefore, to see why early civilisations were both in awe of and feared the mighty Sun. These solar phenomena beckoned natural philosophers, and later scientists, to unlock their true nature and origins.

Making Connections

Modern astronomers who specialise in studying the Sun are still searching through and analysing solar observations made by all manner of instruments across written history. Each observation is a piece of a gigantic puzzle that makes up a more complete picture of our star. With advances in technology came new insights. One of the first major leaps forward was the invention of the telescope. Shortly after the turn of the 17th century, a number of Dutch lens makers worked with two lenses and demonstrated they could be used to magnify views of distant objects. Hans Lippershey often gets the credit as the first to submit a patent to government. Even though the concept and configuration were simple and may have been demonstrated before, the key was that lens making had improved enough to make a useful high-quality magnified image. Spotting something on land or sea out of reach of the sharpest eyes on Earth gave someone with a telescope a distinct advantage in commerce and conflict.

Italian astronomer Galileo Galilei is most widely credited as the first to turn the instrument to the heavens (based on the Lippershey design, but of his own making). English astronomer Thomas Harriot, however, made some observations with a telescope months before Galileo. In any scientific profession, the phrase 'publish or perish' is a potent one. Depending on how good someone is at spreading their knowledge, one person can receive the spotlight over another regardless of who really did what first. Galileo was one such person. I encourage you to keep this in mind when I highlight some of the key figures hailed as champions of solar observations and physics in this chapter. Hidden figures are being uncovered all the time and we navigate a written history that cannot be seen as unbiased, especially in older records.

Those rarely seen and vague 'blemishes' called sunspots, observed pre-telescope by John of Worcester and others, could now be viewed with greater ease by astronomers. The history of who first observed sunspots by telescope is a tangled web of competing desires to be known as 'the first'. By letter, dated observation or publication there are four people often mentioned in this time of early telescopic solar observation: Galileo, Harriot, Johannes Fabricius and Christoph Scheiner. Despite the newfound magnifying power of telescopes, it was often suspected that sunspots were not on the surface of the Sun, but objects close to it instead. Many theorised they could be undiscovered planets or comets passing between the Sun and the Earth, blocking some of the Sun's light.

By combining the telescope and a device called a camera obscura, father and son team David and Johannes Fabricius may have created the first **helioscope** and paved the way for the addition of photographic film in later centuries. This is contested as Christoph Scheiner from Swabia (in modern-day Germany) is often credited with coining the term 'helioscope' and producing this new combination of instruments as well as the use of coloured glass to tame the Sun's rays. Thomas Harriot employed more questionable alternative techniques: he used times when the Sun was low in the sky at sunrise and sunset or when there was a heavy mist or fog so the atmosphere or weather dimmed sunlight for his telescope. Some things never change, as today it can still be equally dangerous observing the Sun. Direct sunlight through a telescope damages your retinas painlessly. In other words it can blind you without you even realising it's happening! So either a sufficient artificial filter or indirect observation is needed to protect your eyesight.

The camera obscura is an evolution of the simple pinhole projection that projects images onto a wall or screen in a darkened space. Instead of just a small hole, a light source, and a long distance between the hole and a screen, a camera obscura includes a lens to shorten the focal distance and magnify the image while preserving a lot of the light passing through it. That way instead of fighting a losing battle to make the distance between a pinhole and a screen greater to produce a bigger but dimmer image, you instead end up with a shorter distance and a clear, bright, magnified image.

In the 17th and 18th centuries, astronomers studying the Sun could use this technology to make sketches of sunspots to see detail and changes in their size and shape. In 1630 Scheiner eventually completed a book called *Rosa Ursina*, which became a comprehensive reference for astronomers interested in sunspots. It came complete with many detailed drawings of observations and recording methods that were used in subsequent centuries. It's worth noting that a pamphlet published by Johannes Fabricius preceded this by 19 years and proposed the most significant aspect of these mysterious features. He worked out that due to their movement and lack of **parallax**, they were features affixed to the Sun, not celestial bodies near it like Mercury and Venus. Sadly his pamphlet was not widely distributed, and his father, David

Fabricius, a pastor and fellow astronomer, did not wish to defy Church teachings that the Sun is perfect and so denounced Johannes' claim.

As lens making improved further and the industrial revolution provided new manufacturing techniques, larger and more powerful telescopes could be crafted. Larger lenses are important when it comes to gathering more light from faint objects like nebulae, or discerning stars that are very close together in the sky. These larger lenses also mean that objects can be magnified more without losing lots of detail. This set the stage for photography with telescopes, which we'll come to later.

Despite the number and size of telescopes growing ever larger in the 17th and 18th centuries across Europe, little more was discovered about sunspots compared to the rapid investigations that had taken place in the first two decades after the introduction of telescopes. The sunspots were eventually confirmed to be on the Sun's surface, their movement proved the Sun rotates on its axis, they were shown to grow and shrink in size and complexity over time, and they each had a lighter outer region and a darker central region. As it turns out, observations of sunspots from the mid-17th century onwards were few and far between owing to two very different reasons. Firstly, the number of sunspots dwindled for a stretch of 70 years or so. Secondly, finding new planets was still the main pursuit for many astronomers. There is speculation that some astronomers may have only observed sunspots to identify candidates for an undiscovered planet moving across the face of the Sun. They disregarded many as they were not suitably round in shape.

New technology and new techniques provided a way forward for solar research, but for real progress to be made, the way of 'doing science' on a larger scale needed to change. From the late 17th century onwards, astronomical observations were no longer gathered by learned natural philosophers working alone, sending their results out into the wild to be debated by other lone investigators. Scientific institutions had been formed (for example, the Royal Society in 1660), observatories with multiple telescopes had been founded (such as the Royal Observatory in Greenwich in 1675) and eventually, in 1833, the term 'scientist' was coined to describe this pursuit as an actual job (by William Whewell, 1794–1866). These factors combined provided fertile ground for the sharing of information and coordination of data gathering on a scale the world had never seen. Though some were late to the party or saw it as a by-product of planet hunting, these types of facilities and structures eventually led to regular, systematic observations of the Sun.

One astronomer who helped focus attention on sunspots despite being a planet hunter himself was German astronomer Samuel Heinrich Schwabe. Over the course of 42 years (1825–67), Schwabe made over 8,000 sunspot drawings. As his sunspot records grew, Schwabe noticed a pattern in the appearance and disappearance of sunspots that seemed to come in a ten-year cycle. While not the very first to suggest a potential pattern, his observations were extensive and piqued the interest of then director of Bern Observatory, Rudolf Wolf. Wolf began a systematic observing programme there in 1847 as well as creating the 'Wolf number'. This was a tally of sunspot numbers gathered from multiple observers and observatories. Thanks to his processes and data gathering, Wolf refined Schwabe's original sunspot cycle to 11 years – a figure still used today. The discovery of this cycle was key to unlocking more of the Sun's secrets and forms the foundation for modern-day solar physics.

Investigations into the nature of light in the 17th, 18th and early 19th centuries by titans of science such as Isaac Newton, Christiaan Huygens, James Gregory, Thomas Young and James Clerk Maxwell revealed light to be far more than just something that illuminates objects and provides warmth. This book could not do justice to the breadth of exploration to which they and others committed their lives, but it's essential, though, to mention that their work vastly deepened the scope of astronomical research. With a greater understanding of the nature of light, astronomy went from the study of positions and appearances of objects in space to an investigation of what they were made of, how they came to be, and what would happen to them.

Not all solar research focused on sunspots. After experimenting with telescopes and camera obscuras for a skin-deep exploration of the Sun, an instrument called a **spectroscope** appeared on the scene. It allowed astronomers to pick apart the Sun just like they could a metal or a mineral in a laboratory on Earth. In its most basic form, a spectroscope uses either a prism or a small grating with many narrow slits to split the light coming from the Sun into its component wavelengths. In 1666 Sir Isaac Newton had discovered that sunlight was made up of a rainbow or spectrum of colours. Fast forward to 1814 and Joseph von Fraunhofer created the first modern spectroscope and used it to better view this spectrum. He noted that his 'rainbow' had very small gaps in it and that some specific shades were missing. These gaps became known as the **Fraunhofer lines**. Something appeared to be blocking parts of sunlight, but not in

31

the traditional sense of a solid object like a planet getting in the way because the rest of the colours that made up the sunlight were still getting through.

A few decades after Fraunhofer's invention, physicist Gustav Kirchhoff and chemist Robert Bunsen experimented with heating elements with Bunsen's now-famous burner, a staple of every school chemistry lab. It had been known for some time that elements in the form of salt-like lithium or calcium thrown into a flame would change their colour, but Kirchhoff and Bunsen viewed the change with a spectroscope of their own invention. While Fraunhofer saw dark lines in a spectrum of colours with his spectroscope, it was now possible to see bright lines of colour with this new setup. Each element had a corresponding spectrum or 'light fingerprint' that could be consistently matched. Kirchhoff and Bunsen realised they could effectively 'fill in' most of the gaps in the solar spectrum with the colour patterns they were seeing for various elements. Though a more complete understanding would come later, there was enough evidence to deduce that the same elements that created bright lines in the lab were responsible for the dark lines in Fraunhofer's spectrum. Now solar physicists had full access to the ingredients of the Sun. However, like puzzle pieces that end up down the back of the couch, some of the Fraunhofer lines didn't have an existing matching element. Only recognised as unique lines in the solar spectrum from observations of the Sun's atmosphere during a solar eclipse in 1868, a hypothetical missing element was named after the Greek god of the Sun, Helios – 'helium'. It would be another 14 years before it was discovered on Earth courtesy of emissions from an eruption of Mount Vesuvius in 1882.

Meanwhile, and in parallel to the leaps and bounds made in the understanding of sunlight, work in the field of magnetism was moving on too. This might seem like a bit of a steer away from anything to do with the Sun, but there is a connection. By the mid-19th century, many major observatories had established magnetic and meteorological departments. A hot topic in the field of navigation was the effect of the magnetic force on the compass, plaguing many seafaring nations. Leading the charge for global collaboration was Irish astronomer and geophysicist Edward Sabine. The collaborative effort to collect enough data to understand and interpret magnetism was known as the Magnetic Crusade and thanks to the practical navigation conundrum, astronomers concerned with the Sun added another string to their bow – magnetic observations. Sabine played a part in connecting magnetic activity with the appearance and disappearance of sunspots. In 1852 he was able to compare the records of geomagnetic activity gathered by the observatory network and show a precise correlation between the 11-year sunspot cycle and an 11-year cycle of **geomagnetic activity** detected on Earth. While a matching number doesn't prove the magnetic activity was linked to sunspots, it did raise a question – what could magnetism have to do with the Sun?

As lines of solar inquiry, interest and technological advancement converged, a global event drove home the connection between activity on the Sun and phenomena seen on Earth. The combined knowledge of all three would be needed to recognise its importance. On 1 September 1859 British astronomer Richard Carrington was viewing an intriguing and complex sunspot group on the Sun with his telescope when flashes like lightning and as bright as the full glare of the Sun appeared among the group. One can only imagine the excitement he felt. As you'd expect, he rushed to get his assistant to confirm what he was seeing, but sadly, within the space of a few minutes, the 'flash' had greatly dimmed. While Carrington compiled his findings, unbeknown to him a huge mass of charged particles from the Sun was closing in on our planet with ferocious speed, taking less than 18 hours to reach its destination. Reports of strange goings-on appeared in newspapers across the world. People as far south as Cuba witnessed a light show in the skies usually reserved for those living nearer the North Pole – the aurora. And telegraph operators found they could no longer transmit and receive successfully, with those in Paris reporting sparks on the wire when they cut their power.

This one event, now known as **the Carrington Event,** encapsulates almost everything that fascinates scientists today when it comes to studying the relationship between the Sun and Earth – from observing an event taking place in a complex sunspot group through to tangible effects experienced on Earth, 150 million kilometres away. We'll return to the contemporary significance of such events and the importance of space weather in the chapter of the same name. For now, in the chronology of humankind's quest to understand our star, the Sun, it's likely this event accelerated the joining of connections and further fueled the study of an already intriguing celestial body.

The last few pages have been littered with names in the fields of astronomy and physics that demonstrate the abundance of rich and interesting stories that are to be told of science and scientists. A surname that surfaces most in solar physics, historic and contemporary, is undoubtedly the name Maunder. In efforts to redirect credit where it is due, I did not give the 70-year period of low sunspot activity its best-known name: the Maunder Minimum. The surname links to not one, but two talented astronomers, Edward Walter Maunder and Annie Maunder (née Dill Russell), who dedicated much of their lives to studying the Sun and sharing their understanding beyond typical academic circles.

In 1874, the Royal Observatory in Greenwich stood as a shining example of consistent sunspot record-keeping. These consisted of daily observations by telescope of the number of sunspots, the area of the solar disc they covered, and their positions. Thanks to contributions from partner observatories in India, South Africa and Mauritius, inclement British weather didn't lead to huge gaps in record-keeping. Eventually, in 1976, the duty of daily sunspot observations was passed on to the United States Air Force and subsequently the US National Oceanic and Atmospheric Administration (NOAA) before funding ceased in 2005. Other telescopes, on Earth and in space, are now almost constantly observing the Sun's activity, helping the world's solar scientists keep an ongoing record of sunspots as well as other features that are part of the Sun's domain.

The consistent record-keeping at Greenwich started with the aforementioned Edward Walter Maunder. Soon after he began work there, he met his match in life and work in the figure of Annie Dill Russell. Annie was employed at the Observatory in 1891 to join a small team of 'lady computers' during the time of the 8th Astronomer Royal, William Christie. For Christie, this was an experiment and an opportunity to employ a group that could go beyond the normal duties of human calculators for astronomical tables. The women were part of the team that embraced what was then a new field and a game-changer for astronomy – astrographic observations. This was the combination of astronomical and photographic techniques. Astronomers no longer had to struggle with tracing projected images, which was prone to error and was limited by the power of the human eye. Now light could be captured and preserved for future records and there was the opportunity to gather consistent observations if the techniques and equipment used were noted and shared. Today astronomers use digital cameras that have great advantages over chemical photography. The chips in digital cameras are more sensitive to incoming light, the images can be investigated using digital techniques, and it's far easier to store digital files on computer drives instead of the huge thick glass plates that were used in the Maunders' time.

I'm sure it hasn't gone unnoticed that Annie is the first female scientist I've mentioned in this book, a reflection of both the gender gap in scientific professions then and now, and the institutions of the past that, deliberately or not, failed to provide equitable opportunities for women as much as men. Annie was a secret superstar

The modern recreation of the entrance to the ancient passage tomb of Newgrange in County Meath, Ireland. First discovered in 1699 and restored between 1962 and 1975. The roof box feature above the entrance aligns with the rising Sun on dates around the winter solstice and illuminates the floor of the inner passage.
INTERFOTO/Alamy Stock Photo

時定方知

The cover of a 19th-century compact sundial from China, which contains a table of sunrise and sunset times for different periods of the year. The inscription on the front translates to 'Secret of the Sun'.
National Maritime Museum, Greenwich, London

A 19th-century educational wall hanging depicting the Sun and planets, from Mercury to Neptune (M to N, left to right). The planets are presented to scale, showing that the diameter of the Sun is 100 times that of Earth. The Sun is depicted with a number of large sunspots, each one with an umbra (dark centre) and penumbra (lighter outer area).

National Maritime Museum, Greenwich, London

Annie Maunder (née Dill Russell, 1868–1947) was a pioneering astrophotographer, astronomer and science writer. With her husband, Walter, Annie carried out pioneering work to record sunspots, but she also undertook outstanding research in her own right. Denied professional status in various circles due to her gender, she worked tirelessly to further our understanding of the Sun, later exploring ancient astronomy and sharing her findings with the public.

National Portrait Gallery, London

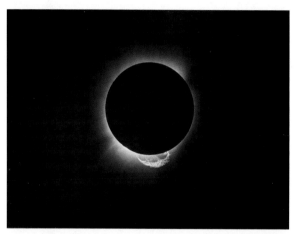

Scan of a photographic plate capturing the 1919 total solar eclipse. As well as capturing the ever-present glow of the solar corona, the Sun put on a show here with an enormous prominence – a loop of plasma rising out of the Sun and extending substantially into the corona. Until the invention of the coronagraph in the 1930s, total solar eclipses provided the only opportunity for astronomers to observe the Sun's outer atmosphere.
National Maritime Museum, Greenwich, London

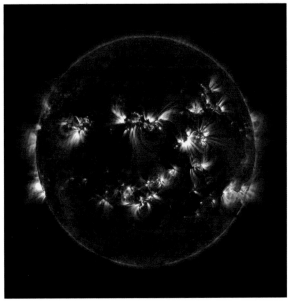

An image of the Sun captured by NASA's Solar Dynamics Observatory (SDO) in 2015 at the height of the previous solar cycle. The cameras on board the space-based observatory capture ultraviolet light, giving the best views of bright, active regions where loops of super-hot gas rise and fall. They are governed by magnetic fields and will erupt under the right conditions. *Solar Dynamics Observatory, NASA*

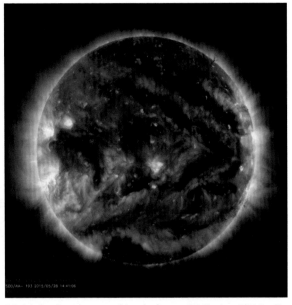

SDO/AIA 193 2015/05/28 14:41:06

Multiple images combined from SDO in 2015. By viewing different parts of the ultraviolet section of the electromagnetic spectrum, features of different temperatures can be seen. Giant, dark filaments are visible on the right half of the Sun and appear to form an arrowhead. When this type of feature rotates to the edge of the Sun from our perspective they will be seen as prominences (i.e. loops).

NASA/SDO

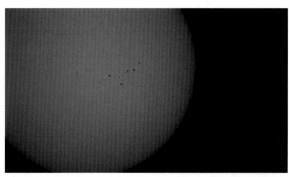

White light image of the Sun captured using a DSLR camera and 130 mm Newtonian telescope fitted with solar film. The image features a complex sunspot group. Penumbra and umbra features are visible on the largest sunspots. Captured on 26 April 2021 from Dublin, Ireland.

Brendan Owens

in the relatively early days of astronomy photography. After defying the restrictions of the time, passing all exams at Girton College, Cambridge, but not receiving the degree she had rightfully earned, she worked under Edward Walter Maunder in the newly created Astrographic Department, and became quite adept at using the photoheliograph to capture the disc of the Sun daily.

The appointment of a handful of 'lady computers' in the late 19th century was sadly an anomaly for the time. The idea of employing more women at the Royal Observatory was abandoned near the turn of the 20th century and only resurrected during the Second World War out of necessity rather than as part of a campaign for women. Despite resigning shortly before marrying Walter, Annie continued to pursue solar observations as a member of the British Astronomical Association (BAA). She later returned to the Observatory as a volunteer to continue photographing the Sun. As part of eclipse expeditions across the globe with the BAA, Annie became widely known as an expert in photographing the outer atmosphere of the Sun – the corona. Those lucky enough to see totality with their own eyes could make this out as a ghostly glow around the black disc of the Moon. Her customised combinations of portable telescopes and cameras put some of the greatest observatories in the world to shame. Annie's photography of solar eclipses also allowed for a better view of bright arcs called **prominences**, which extend out of the Sun and back to its surface. Unfortunately, despite gaining entry immediately to the BAA, she was not only denied entry to the Royal Astronomical Society of the time but also her fantastic images were credited Mrs Walter Maunder.

Combining their knowledge and skills, Annie and Walter worked long and hard on cataloguing sunspot observations together. And while Mr Maunder may have received the majority of the credit and had the opportunity to speak with authority where Mrs Maunder couldn't, their collaboration was true teamwork. Their combination of efforts and records gave rise to a hand-drawn diagram known as the Butterfly Diagram (see Figure 3). In it, you can see the position of sunspots by solar latitude and the area they take up on the face of the

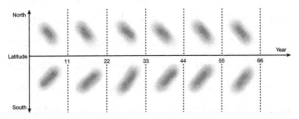

Figure 3: A modern, simplified representation of the famous Butterfly Diagram that was originally hand-drawn by the Maunders. By graphing the position, size and number of sunspots over years of daily solar observations, Annie and Walter visualised this repeating 11-year pattern that gave the diagram its name.

Sun. It's easy to see how it got the name. It's a visual representation of the solar cycle that had been discovered decades earlier by Schwabe and Wolf. This diagram is still being added to as one (mostly) consistent record of sunspot activity. The Maunders' interest in sunspots and particularly Walter's exploration of the 70-year near-absence of sunspots previously credited to fellow astronomer Gustav Spörer led to the naming of the period in his honour. Nowadays it's easier to find a reference online to the Maunder Minimum than either Annie or Walter. Annie's profile, however, has been raised in recent years. A medal bearing her name is awarded by the Royal Astronomical Society for outstanding contributions to outreach and public engagement, and a new telescope was named after her at her former working place, the Royal Observatory.

To wrap up this exploration of the building blocks that form the foundation of our modern-day understanding of the Sun, let's look at one last solar pioneer: the eminent American astronomer and visionary George Ellery Hale. Hale founded the Mount Wilson Observatory in 1904 in Los Angeles and, as director, was responsible for the construction of some truly gigantic telescopes, even by today's standards. From the towering 60-foot (18 metres) high Snow Solar Telescope through to the gargantuan 200-inch (5 metres) wide Hale Telescope completed after he passed away, Hale expanded the scope and scale of astronomy in general and solar physics in particular.

Hale went to work immediately with the increased resolving power of the mammoth Snow Solar Telescope combined with a **spectrograph**. This instrument took photographs of the solar spectrum, which could then be closely studied. The splitting of sunlight into its component colours was, as explained before, the basis for spectroscopy and enables the study of light 'fingerprints' for the elements in the Sun and other stars. These fingerprints are made up of multiple discrete colours and each one of those corresponds to a wavelength and frequency of light on a broader spectrum known as the **electromagnetic spectrum**. We'll hear more about that in the next chapter. Using a modern understanding of the connection between temperature and wavelength, Hale was able to show that sunspots were cooler than the rest of the Sun's surface by a few thousand degrees.

Hale also knew that Dutch physicist Pieter Zeeman had discovered that magnets could have a visible effect on a light spectrum. The Zeeman Effect, as it became known, showed that lines in a spectrum were split in the presence of a strong **magnetic field**. Zeeman found this out in a lab setting with a sodium flame and an electromagnet and hinted in his paper that the effect could be of great importance for studies of objects in space. Initially Hale couldn't see anything in the Fraunhofer lines of the solar spectrum. At the time he was studying calcium lines where no splitting effect was seen in his photographs. However, in 1907, photographic plates more sensitive to the red end of the light spectrum were created, giving better access to the spectral lines of hydrogen.

Thanks to this improvement in photography, Hale was able to see a noticeable split in the hydrogen lines and so he found the first definite link between sunspot activity and a magnetic influence. With this he made one of the most groundbreaking discoveries about our Sun: it is magnetic in nature. While decades earlier Edward Sabine had shown geomagnetic effects correlated with the solar cycle, this was a direct observation of magnetic effects in the Sun. This finally connected activity experienced during periods of high solar activity here on Earth, such as extreme Northern and Southern Lights and radio disruptions, with the ever-changing structure and number of sunspots.

In 1924 Hale invented a new type of instrument called a **spectrohelioscope** to study sunspots in a new way. This instrument was used to look at very particular wavelengths of light from the Sun. In one of the quirkier aspects of physics, light can be seen acting as both a wave and a particle. As a particle, light can be understood as packets of energy. With Hale's instrument just one flavour of these packets of energy can be isolated and studied. By ignoring the soup that is the combination of all the packet flavours from the elemental ingredients of the Sun, astronomers can concentrate on activity occurring at a particular temperature. With this invention, Hale and others could see, without difficulty, refined detail in not just sunspot structure, but also the structure of the relatively quiet parts of the Sun where no sunspots are active. Features called granules give a textured look to the surface of the Sun that resembles crocodile skin. Hale later improved the instrument by including photography to create a spectroheliograph. Due to this adaptation, images could be captured and preserved to be analysed later.

Now the stage is set for solar physics as we know it today. By the mid-20th century, astronomers could split visible sunlight with ease and accuracy, match spectral lines to elements, measure areas of magnetic strength on the Sun, and regularly photograph and monitor solar activity in visible parts of the light spectrum. The next chapter breaks away from the traditional stereotype of astronomers peering through a long telescope tube, as we delve into what makes up the invisible Sun.

The Invisible Sun

To get a deeper understanding of the Sun and what's going on below its surface we need to explore the light that our eyes cannot see as well as rely on knowledge of matter from labs on Earth. In this chapter, we'll break away from a chronology of discoveries to explain the features of the Sun from its core outwards to the known limit of its influence in the Solar System. Get ready for a trip into the heart of the Sun.

Some early astronomers believed that the Sun had an Earth-like surface, with a luminous atmosphere whose sunspots were seen as holes to peer down into the cool surface below. Some even suggested the

Sun was habitable! Today we understand the Sun to be an immense ball of super-hot gas – a state of matter known as **plasma**. Some of the ideas explored in the rest of this book, particularly when it comes to interactions between the Sun and Earth, require a bit more information about what plasma is and how it acts.

When a solid like ice is melted it becomes a liquid. If you heat the liquid enough, it becomes a gas. If you give that gas enough energy, it doesn't just become a very, very hot gas, its state of being changes entirely. Atoms are no longer able to hold onto all their **electrons**, and so we end up with a hot soup of some freely moving electrons and separate **nuclei** of atoms called ions. The various configurations of atoms, the elements, that we find in the Sun are hydrogen (71 per cent) and helium (27 per cent). The rest (2 per cent) is made up of a sprinkling of other elements.

Freely moving electrons are responsible for a lot of things. Old, box-shaped cathode-ray tube (CRT) televisions for instance have a tube inside. Electrons are fired through it onto a screen repeatedly to make moving images. If you've ever wondered how YouTube got its name, now you know! Moving electrons are in fact the basis of our modern world. The transfer of electricity is basically a carousel of electrons going from a power source through the machines we need.

In the natural world, or when humans harness them for power, free electrons are subject to magnetic fields. Fundamentally, electrons are negatively charged particles – they can be pushed and pulled with the introduction of magnets. Just to make things even more mind-bending, charged particles that move generate their own naturally occurring magnetic field. As we'll see later, these fields can interact with each other and the magnetic sparring matches that ensue often end in explosive events on the Sun. Consider for now that we're talking about the entire Sun being a ball containing mostly moving charged particles and you might start to see why magnetic field interactions are the culprit for so much of the Sun's daily routine.

With the combined knowledge of what we can and can't currently see of the Sun's workings, solar physicists have been building up the internal blueprint for a star. Going from the centre outwards, the layers of the Sun and in turn other stars are still considered to be: the core, the radiative zone, the convection zone, the photosphere, the chromosphere and the corona (see Figure 4).

If we want to delve straight into the centre of our Sun, we start to rely extremely heavily on the theoretical blueprint of a star rather than investigating goings-on directly. We'll need an understanding of gravitational theory, gas laws and

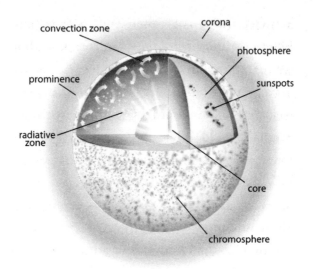

Figure 4: A cross-section of the Sun with major features labelled. These include the various regions of the Sun from the core at the centre where nuclear fusion takes place, outward to the radiative zone, into the convection zone, to the photosphere (effectively the Sun's surface). Up from there lies the lower solar atmosphere, called the chromosphere, up into the outer atmosphere, known as the corona. Sunspots can be seen in the cross-section labelled photosphere. Loops of gas are seen on the left-hand edge of the Sun, which are labelled prominences.

a proper idea of the scale of the Sun. The Sun measures 1,392,000 kilometres across, or approximately the combined widths of 109 planet Earths. It also has a mass of approximately 2,000 billion billion billion kilograms in total. While we can maybe stretch our brains to try and grasp the size of the Sun, its mass (the amount of matter it's made up of) is pretty incomprehensible. A comparison that might help is that the Sun makes up 99.8 per cent of the mass of the entire Solar System. Take a moment to think about that. That 0.2 per cent is made up of the entire planet we live on, the Moon, all the other planets, all the other moons, all the dwarf planets discovered and undiscovered, every possible asteroid, comet, or icy rocky body – combined. 0.2 per cent. The rest of the mass of the Solar System resides in our star, the Sun.

Consider as well that according to our current understanding of physics, everything that has mass experiences a gravitational force. The more mass you have, the greater the gravitational attraction. With 2,000 billion billion

billion kilograms of gas, the gravitational pressure pulling inwards is immense. It almost seems incredible that it hasn't just been squished into itself. As it turns out, another force is balancing our Sun to stop it doing just that – thermal pressure from something going on at the heart of our star.

The core of the Sun in the current blueprint for a star is stated as making up 25 per cent of the Sun's volume. In this region, the immense pressure on the Sun's plasma means temperatures skyrocket to an incredible 15 million degrees Celsius. It is in this nuclear furnace that hydrogen nuclei are able to overcome the **electrostatic force** that keeps charged particles apart from one another in normal circumstances. By doing this, a reaction takes place that is associated with an equation printed on t-shirts the world over: **$E=mc^2$**. Such a tiny equation packs an enormous punch. In very basic terms, it says that an amount of mass (m) in kilograms multiplied by the speed of light (c) in metres per second, squared, is equal to energy (E) in Joules.

During this nuclear fusion of hydrogen, an incredibly small amount of mass is converted into energy according to Einstein's famous equation. Plugging in figures for one of these fusion reactions might not seem that impressive. The amount of mass converted that appears to 'go missing' is incredibly tiny. Even though the value of the speed of light is 300 million metres per second, a single fusion reaction creates a relatively pitiful amount of energy. However, this one reaction at the atomic scale is multiplied billions of times every single second in the core. To put that in context, the Sun is producing 50 million times the annual nuclear power generated in the United States ... every second!

The energy output is only part of the nuclear fusion story, as what remains after each reaction are various particles including a different element – helium. Through these nuclear fusion reactions, the Sun converts hydrogen into helium, energy, **anti-electrons** and some ghostly particles called **neutrinos**. It's thanks in part to these neutrino particles, which can pass through the Sun unimpeded, that we understand more about the physics inside the core. The other particles go into different chain reactions to keep the temperature high so that the fusion reactions can continue ... so long as the fuel is there.

All this means is that the Sun is using up hydrogen, and the rate of loss can seem alarmingly high without giving some additional context. Each second the Sun fuses 400 billion kilograms of hydrogen, which is the equivalent of half the mass of Mount Everest, every single second. However, as mentioned before, the Sun is massive in the true sense of the word. And thanks to all the combined knowledge so far we can divide the mass converted per second into the mass of hydrogen currently in the Sun's core volume to realise that it will take over 4 billion years for the Sun to expend it. For now, the Sun is in a balanced state where the thermal pressure from nuclear fusion is pushing out at a rate equal to the pressure of gravity pulling inwards.

The energy released by the nuclear reactions in the core is in the form of light, but not the visible light we normally associate with the Sun day to day. Instead, it is in the form of gamma radiation. Rather than making us all turn into the Incredible Hulk, this is an incredibly damaging form of electromagnetic radiation for any living beings and the highest energy light in the Universe. Luckily this energy does not have a straightforward way out of the Sun's core and its many interactions through the rest of the Sun's structure to the surface reduce the energy of the end-product that eventually makes its way to our planet.

The journey of a **photon** is not easy to map out. In the core it may seem logical that the crowded nature of that part of the Sun would mean a photon would bump into many particles regularly, but as it's a high-energy photon it has less chance of interacting. We see the ease or difficulty with which electromagnetic radiation interacts with matter when we, for example, consider a broken bone or the internet. If you go to have an x-ray, the nurse or doctor will remain behind a lead-lined screen, such is the ease with which x-rays can pass through matter. Compare that to the struggle to get a good WiFi signal where radio waves from the router are blocked sometimes by just a door or a relatively thin wall.

Moving into the next level up from the core, the radiative zone, the photon will have lost some energy and may be able to interact with particles in this part of the Sun more easily, but now it is less crowded than the core. Here or on any part of its journey, the photon may be absorbed by an atom, with a better chance of absorption in the cooler outer regions of the Sun where more whole atoms exist. This will contribute to the heating of the Sun, with a photon donating some of its energy to the process before being emitted at a longer wavelength to continue its journey. A final zone the photon must seek passage through is the convection zone. Here, again, the density of the area has changed and the photon will once more lose some energy through absorption and emission by atoms. It will finally emerge in its lower-energy form of light from the Sun's surface – the region known as the photosphere.

The question is how long did the journey for our photon take altogether? Without a better understanding of each layer's exact makeup at each moment and the angle photons get sent off in, the estimate for a photon's journey time from the core to the photosphere varies. Most modern computer models of the Sun's interior that are used to calculate a photon's 'random walk' come up with a figure upwards of 100,000 years. This means that despite the eight-minute travel time of light from the photosphere to Earth's surface, the sunlight that strikes your face today could have started its journey back when the oldest known human-made structures on our planet were being built.

It's worth noting that not all the photons of light emerge in the same part of the electromagnetic spectrum. Some will be in the form of light that our eyes can detect while others will emerge as x-ray, ultraviolet (UV) and infrared, radio or microwave light. To know how we detect them requires a little step back in time.

In 1800 William Herschel discovered that when splitting sunlight into its spectrum of colours and recording each colour's temperature, a temperature slightly above room temperature was registered when measuring just beyond the red part of the spectrum. He had broken into the realm of the invisible. Despite incorrectly thinking the Sun had a habitable surface, he did at least give us the key to unlocking more of its secrets.

Modern digital cameras that are sensitive to light beyond the visible part of the spectrum can be manufactured, yet the use of this technology had to go a step further to really open our eyes (safely) to all the Sun's light. Hale and others understood the need to reach higher locations on Earth to get the clearest view of the Universe, though Earth's atmosphere still blocks us from harmful x-ray and UV light from the Sun. Some parts of the UV spectrum still make it through, hence the need for sunscreen to protect our skin. However, as astronomers wanted to investigate all forms of light, cameras had to be launched on spacecraft so that they experienced no atmospheric interference whatsoever.

Today there is a fleet of space-based observatories, operated remotely from Earth, many of which are monitoring the Sun across multiple wavelengths of light on a near 24/7 basis. These spacecraft from NASA, the European Space Agency (ESA) and the Japan Aerospace Exploration Agency (JAXA) specialise in capturing a wide variety of data that not only includes light but also the detection of charged particles and magnetic readings from the Sun. These offer solar physicists data from the Sun's surface outwards.

This still leaves us with a fairly big gap in our data between the theory of the Sun's internal structure and the light that we see emerge from the photosphere. That said, there is a way of indirectly gathering information about the Sun's internal structure beneath the photosphere. On Earth when studying the origins of earthquakes, geophysicists use seismometers to detect vibrations felt at the surface to trace the event back to its source. Even though the Sun is not solid, plasma is a medium and so it vibrates too, and lucky for us those vibrations or oscillations can be seen in the Sun's light. It is a complex endeavour, but to best explain it we can consider a band playing on a stage. Depending on how popular the band is and the size of the venue, the sound waves from their amps will reflect off the wall of the venue and interfere with each other. If we could watch the walls of the venue very carefully, we'd see them shake back and forth. The rhythm and strength of the vibrations will be different depending on whether or not it's a sold-out show, the movement of the audience, and anything else going on between the source (band) and the surface of the walls.

The same can be said for the Sun. The plasma is not static, and as seen by the astronomers of old, the Sun rotates. Also, how tightly packed the Sun is changes from core to surface. The study of the Sun using oscillations seen in sequences of images of the Sun's surface is called **helioseismology**. As you might imagine, things get a lot more complex than at your everyday music venue. The interactions within the Sun can mean the existence of millions of 'modes' of vibrations. So by no means does it give us a complete peek inside the Sun, but some sense of the Sun's shallower interior density and movement can be extracted from the many modes. In fact, this can be done so effectively with modern techniques that it allows solar physicists to study the opposite side of the Sun before it rotates into view as seen from Earth. In this way a rough idea of sunspot activity can be monitored across a full rotation of the Sun. Using helioseismology also allows solar physicists to see the emergence of new magnetic features in basic ways before they emerge as sunspots.

Though not complete, we have a reasonable sketch of the Sun's blueprint to work from. Nuclear fusion in the core generates energy in the form of gamma radiation that passes into the radiative zone where energy is transferred via radiation, as plainly as the name suggests, before entering the convection zone. All the while, some energy is continuing to escape outwards in the form of lower-energy light. It is in this zone that some more traditional physics are in play. As the plasma moves up through this zone it can be visualised as a boiling pot of water. You may remember that the invention of the spectroheliograph and later solar filters allowed astronomers to view granules that seemed to bubble up from within the Sun. Just like the water in the pot, the hot plasma rises, cools a little, and falls back down before being heated enough to rise and so on. This continuous churning of plasma also means lots of charged particles moving and generating magnetic fields. It's in the convection zone that footprints of sunspots are born. Once the magnetic fields are complex enough, plasma that rises to the surface can become temporarily trapped before the field lines braid and twist enough to snap into new positions and in turn release energy.

The features that result from all this magnetic activity, once hidden to the human eye, now reveal themselves in UV wavelengths. No longer are these simply mysterious voids of black on the surface of the Sun, they are the base of gargantuan ribbons of plasma extending through the chromosphere, curved by intense magnetic field lines back down to another sunspot base to enter the photosphere once more. These features are called coronal loops and centuries before could only be inferred from larger loops of plasma called prominences. Those could be seen occasionally on the edge of the Sun's disc, but only then during solar eclipses when the rest of the Sun's dazzling light was obscured by the Moon. The coronal loops draw energy away from the connected surface of the Sun, therefore cooling it by about 2,000 degrees Celsius and in turn giving sunspots their dark appearance compared to the 5,500 degrees Celsius of the photosphere. Features such as prominences and vast explosions from the Sun called **coronal mass ejections** (CMEs) were photographed in the 19th century, but only during those rare total solar eclipse events.

From the 1930s onwards opaque discs of material were used with telescopes to create false solar eclipses to get a better view of the Sun's outer atmosphere. Astronomers no longer needed to wait around for a solar eclipse to occur. These discs are called coronagraphs and suffered from glare issues on Earth because of the scattering of sunlight in our atmosphere. Once they were fitted to space telescopes like the NASA SOHO spacecraft, however, this wasn't a problem anymore.

It's thanks to these views that we start to get a better understanding of the true domain of the Sun. While many magnetic field lines are closed off, starting and ending on the photosphere at sunspots, others are open field lines that cast out charged particles like protons and electrons in all directions away from the Sun, deep into the Solar System. All the while, the Sun is rotating and creating holes from which a **solar wind** can freely stream charged particles at high velocities, up to 800 kilometres per second. The extent of the Sun's magnetic influence is so vast that it creates a gargantuan magnetic field that engulfs the enormous scale of the entire Solar System.

Its true extent has only recently been investigated by our furthest robotic explorers, *Voyager 1* and *Voyager 2*. These spacecraft were launched from Earth in 1977 and *Voyager 1* (as of writing the first edition of this book) has travelled more than 22.5 million kilometres in that time. A milestone in its journey was reaching a point called the heliopause. The bubble of the huge magnetic field mentioned a moment ago is known as the heliosphere and the heliopause is the point at which the solar wind can no longer push against the stellar winds coming from other stars outside. This milestone was reached by *Voyager 1* in August 2012 – 18 million kilometres from the Sun. It is only here that the solar wind's domination of our Solar System comes to an end.

The immense power and reach of the Sun are staggering and yet on a typical day on Earth its influence on our planet and on our lives is not immediately visible beyond the obvious light and warmth it provides us. To understand the real significance of the Sun on Earth and humankind we need to delve into space weather forecasting.

Space Weather

In the 19th century, as astronomers were starting to make solid connections between the Sun's activity and impacts seen on Earth, many of the effects were either a minor inconvenience or more often they were simply a source of pure wonder and amazement. By the 21st century, the electromagnetic effects experienced when the Sun decides to unleash a stellar belch are felt far more deeply. Today we are connected the world over via our power grids and telecommunication networks. Without them, economies in developed and developing countries would cease to function in their current state, our lines of communication would be cut or greatly reduced, and access to huge amounts of data would be eliminated. It sounds like the plot of a spy thriller where the bad guys use an electromagnetic pulse (EMP) to send civilisation back centuries, but while total global devastation is not on the cards from a singular massive solar storm, major disruption for whole countries or continents for months at the very least is not such a wild idea.

The Sun has given us warning signs in the past such as the Carrington Event of 1859, but at that time the only major electrical infrastructure that could be affected was the telegraph system for a few countries in the world. Today we have numerous power stations, radio transmitters and receivers, satellites and more – all vulnerable to geomagnetic effects generated during high periods of solar activity. This means that there is more than just a passing interest in trying to forecast space weather.

Governments globally are taking measures to safeguard technology and many have active space weather forecasting initiatives linked to data gathering instruments on and off the planet.

To understand the Sun–Earth system let's travel 150 million kilometres once again to our star, the Sun, and spend some time with a solar storm from birth to impact on Earth. In our fictionalised (but realistic) tale, we start with a group of sunspots, perhaps small in number and low in complexity to begin with. Over the course of a few days, the patterns of magnetic activity become more complex, loops of magnetic field lines shrink and grow. Some intertwine until, in just a few brief moments, they 'snap', unleashing pent-up energy with ferocious effect. Part of the release is witnessed as a flare – a blast of high energy light in the x-ray part of the electromagnetic spectrum. In a mere eight minutes, the x-ray energy arrives at Earth and heats the atmosphere enough to change its density profile. This interferes with both short-range radio communications on Earth and satellites in low-Earth orbit. In extreme cases, such an atmospheric change can increase the drag on satellites and unless addressed quickly, the satellite can slow down, tumble and eventually burn up. Hundreds of millions of dollars' worth of satellites have already been lost to date because of this effect.

While the impact of flares might seem a few steps removed from affecting humans, there are direct consequences. This x-ray energy is a form of radiation that is damaging to the human body and, beyond the occasional medical x-ray, we should not be exposed to it. Most at risk are the astronauts currently working in space, although the walls of spacecraft, such as the International Space Station, are designed to protect from even the worst solar radiation. Closer to home, high-altitude flights are also exposed to more radiation than someone on the Earth's surface due to the decreasing absorption powers of our atmosphere with altitude. However, even in this case any passengers at cruising altitude at that moment will simply get the equivalent of a simultaneous medical x-ray.

In the meantime, our flare was just the first wave of an unintentional solar attack. The solar flare imparts energy to charged particles already streaming towards us in the form of the solar wind, accelerating them to reach Earth faster than normal. Almost simultaneously with our flare, a huge swathe of the Sun's matter is released into space. This is a coronal mass ejection and is sometimes referred to by itself as a solar storm. A billion tonnes of matter from the Sun is blasted away and carried by the solar wind off into the wider Solar System. In our tale, the trajectory of the blast is in the direction of Earth so that we get the full brunt of the impact. If the Sun had spun around a little more or a little less before belching, the CME may have brushed past us, maybe hitting distant Mars instead. Not all CMEs happen alongside a solar flare, but many do – particularly in the most energetic events.

It takes the CME much longer than the flare to reach us; they typically take between one and three days to make it to Earth. Here we are protected to a certain extent by our magnetic field as the first line of defence, followed by our atmosphere. However, intense bombardment by a mass of charged particles like protons and electrons in the CME can greatly deform Earth's magnetic field. This means it temporarily crumples and allows more charged particles than normal to travel along Earth's field lines to enter our atmosphere in and around the magnetic north and south poles.

Under normal solar wind conditions at the extreme latitudes of the Earth, we see the Northern and Southern Lights, also known as the *Aurora Borealis* and the *Aurora Australis*, respectively. Curtains of green, red or purple light can be seen dancing and shimmering in the sky and their appearance is truly mesmerising. They're caused by the charged particles from the Sun striking atoms of oxygen and nitrogen, between 80 and 300 kilometres above the ground. By doing this, they excite the atoms temporarily and when they relax they release photons of light that we can see. Their colours depend on the energy of the interactions and the element involved. It's a nice bookend to the discovery of the Sun's elements through spectroscopy and the same concept is involved here.

In our tale of solar bombardment, imagining this to be a massive CME striking the Earth, the deformed magnetic field of the planet would mean the aurora could be seen on the night-time side of the Earth close to the equator, as has been recorded during solar storms in the past. However, there are other, more concerning effects afoot.

The CME causes the Earth's magnetic field to change dramatically in a short period of time and this induces an electric current through the Earth. This phenomenon is known as a geomagnetically induced current (GIC). A small experiment in which a magnet is spun in a coil of wire can show this in action. If the coil is hooked up to a device to measure electric current, the simple action of spinning the magnet in the coil registers a small current in the measuring device. Now imagine this on a much larger scale. The GICs generated by a CME striking Earth can induce huge currents in any long conductors on Earth, including pipelines, railway tracks and electricity cables. In the latter case such an event can completely overwhelm power stations. A famous example of this occurred in Quebec, Canada, in 1989. The geomagnetic storm caused by a CME that erupted from the Sun three days earlier induced a ground current so strong it resulted in a blackout that lasted 12 hours and affected millions of people. This was a real wake-up call for the world to pay attention to monitoring systems for power grids the world over. It's estimated that the 1859 Carrington Event resulted in geomagnetic storms that were ten times more powerful.

There are a variety of space weather effects which come under the term 'disruption of communications'. As mentioned previously, satellites, a cornerstone of modern telecommunications, are vulnerable. As well as being potentially deorbited by changes in the atmosphere, their way of communicating would also be disrupted:

most satellites communicate with ultra-high frequency radio transmissions back and forth from Earth, so an increase in the amount of plasma present in the Earth's upper atmosphere would cause disruption to these. The interaction between charged particles from CMEs and the atmosphere can also interfere with more earthbound transmitters and receivers that rely on using the **ionosphere** to bounce signals back down to Earth over long distances.

On top of the satellite transmission issue, CMEs consist of dense clumps of charged particles and this has a direct effect on satellite hardware. Many of us have experienced the phenomenon where we hear crackles when taking a piece of clothing off, particularly if you have long, dry hair. In the dark you might even see a few sparks. This is a discharge of static electricity where friction between appropriate materials strips electrons from one surface, transferring them to another. Something a little more dynamic happens in the case of satellites impacted by charged particles. If one surface accumulates more charge from a solar storm than it can hold, it looks to discharge to the next viable surface. Often this is on the satellite itself resulting in a spark that can cause false readings, or, even worse, damage a critical component that will break the satellite permanently.

While all of the above may be the extreme version of space weather, it's all possible and examples are available from recent history. At the time of writing the Sun is just waking up from the last lull in its solar cycle, and we are officially in Solar Cycle 25 of recorded solar activity. The last maximum was not a particularly high peak in solar activity compared to previous cycles. The next few years will reveal the extent of the next maximum.

So, with some time potentially on our hands, what can be done to protect against the numerous effects of space weather? Well, for starters, great strides have been made in the past decade in consistently monitoring the Sun and the Earth–Sun environment. The fleet of space observatories mentioned in the previous chapter is equipped with more than just telescopes and cameras. They have instruments to detect the strength and density of the solar wind and magnetometers to detect fluctuations in the Earth's magnetic field. These types of instruments are good for understanding active effects, but, to prepare for the worst, solar physicists are trying to improve their understanding of sunspot development and what configuration of magnetic footprints is likely to produce a flare or CME event. To come up with a recipe book for these events, huge amounts of data across multiple wavelengths need to be analysed and combined with an understanding of magnetohydrodynamics in particular. It's a field of physics with a complex name that hints at its complex nature. If we break it down, we have 'magneto' as in magnetism, 'hydro' as in fluids, and 'dynamics' is what it is – the complex, ever-changing interactions of fluids and magnetic fields.

Spacecraft are funneling huge amounts of data back to Earth, but there is so much that some scientists have enlisted the help of citizen scientists to comb through data by eye to recognise shifting patterns and particular features. The Zooniverse projects Solar Stormwatch and Sunspotter are great examples of the human knack for pattern recognition and participation required no prior knowledge. In fact, prior knowledge can be a disadvantage. Knowing what recognising something means can unintentionally bias the user to force a particular outcome. With large sample sizes viewed by thousands of users, these types of projects provide a way forward for training computer programmes to better recognise features linked to the beginnings of solar storms.

Keeping relevant industries updated on the latest predictions means they're able to spring into action, in the event that a CME occurs, there is normally at least a day's lead-time to make necessary preparations. In the case of a power station, transformers and other power systems can be temporarily taken offline with a major, but short-term, impact for customers. So rather than frying components that could take months to replace, they switch them back on a few hours or days later unharmed. For both big business and our personal lives, this changes the level of impact felt and could greatly reduce recovery times.

Accurate space weather forecasting might appear to be the pinnacle in our understanding of the Sun, but our journey of discovery is not at an end. There are some very big solar secrets yet to be revealed and that's what the last chapter of this book intends to cover.

Big Unanswered Questions

Science, in general, boils down to asking questions and finding answers. Many answers that are uncovered through varied and repeated scientific investigations often create more questions. Those new questions can refine our understanding of a murky phenomenon or, even more excitingly, they can completely upend current theories.

The solar investigative tools we have to hand are impressive. Throughout this book, we've seen record-keeping and data collection from near and far with technology that allows us to go well beyond what humans alone can capture.

Due to their vast distance from Earth, most stars appear as distant pinpricks of light, even with our most powerful telescopes. However, the Sun's relatively short distance from Earth allows humans to study it up close and in detail. That said, there are still some major mysteries we have yet to unlock, from a temperature conundrum that contradicts our most basic human knowledge to a controversial connection to the climate on Earth. For each mystery, we only have some clues and an exciting voyage of discovery awaits those who take up the mantle of solar scientist.

The coronal heating problem

Since humans first made fire, there came some fundamental and instinctual knowledge: the further we are from a fire, the cooler it gets. The Sun throws that notion out the window. With a core temperature of 15 million degrees Celsius, the temperature of the Sun drops off as you go upwards through the radiative zone, convection zone, and reach the photosphere – the so-called surface of the Sun, coming in at 5,500 degrees Celsius. Bizarrely, beyond this and the next layer up (the chromosphere), the temperature of the outer atmosphere of the Sun, the corona, rockets to a whopping 2 million degrees Celsius and higher. This is the equivalent of people close to the fire feeling warm while those further away are incinerated!

So, what's going on? Well, the topic wouldn't be in this chapter if there weren't a few unconfirmed ideas. They all trace the mystery back to the problem of energy transfer. Each zone up from the core is named after a method of heat transfer – radiation, then convection. Beyond the photosphere the energy transfer method becomes blurry. There are three well-known candidates, but none is a clear winner.

As mentioned in the previous chapter, areas of sunspot activity are the base of huge arcs of plasma called coronal loops. When the magnetic fields that govern these features braid and snap, they experience an energetic outburst observed as a solar flare, CME or both. With the latest high-resolution images of the Sun captured in UV, astronomers have seen miniature versions of these energetic outbursts within coronal loops. They are much smaller in scale but numerous and give the impression of sparkles in the arcs of plasma. Astronomers have called them 'nanoflares' since they were predicted in the seventies and more frequent observations of nanoflare candidates are being recorded with the latest imaging technologies. At the time of writing very recent close-up observations of a suspected nanoflare have shown that they do superheat the plasma in their immediate vicinity to millions of degrees Celsius. Despite new detailed observations, there is still a very big question mark over the number of these predicted events. Even with the snap, crackle and pop of many of these features at any given time, there may not be enough to account for the energy transfer needed to match the corona temperatures measured.

Next up is a solar feature that hasn't been mentioned yet called a spicule. These are brief jets of plasma hundreds of kilometres wide that shoot up from the photosphere for 15 minutes at a time. Their origins are still debated, but as there are many of them on the Sun at any given time, it has been suggested they could bridge the gap between the photosphere and chromosphere for energy transfer to the corona. Unfortunately, their combined estimated energy output falls well below what is needed to address the coronal heating problem.

Finally, there is a suggested energy transfer that doesn't start its journey close to the surface of the Sun at all and instead is theorised to start at the core and ripple outwards. As we saw in relation to helioseismology, pressure waves can travel through the Sun's plasma medium, rippling their way out and interfering with one another in different ways. This helps solar physicists understand a bit more of what happens beneath the photosphere. There are other waves called Alfvén waves that some scientists say can interact with plasma and trigger energy transfers up into the chromosphere and corona. The physics behind these interactions is complex and not entirely known and direct observational evidence is lacking. It's hoped that **NASA's Parker Solar Probe** will be able to put Alfvén waves and nanoflare theories to the test with its up close and personal investigations of the Sun's atmosphere.

Magnetic origin story

Magnetism and magnetic activity plays a fundamental role in the life of our star and all others, but so far we've looked at lots of localised bits of magnetic activity. However, if you cast your mind back to the solar cycle and particularly the Maunders' Butterfly Diagram (page 43), these point to some larger overall mechanism at work. At the start of each solar cycle, sunspots grow in number and start appearing at latitudes far away from the equator nearer the poles. As the cycle progresses, the sunspot appearances occur closer and closer to the solar equator before diminishing. This cycle of moving magnetic activity is the tell-tale sign that a sun-wide transformation is taking place – the magnetic north and south poles are flipping. This 22-year cycle of the north pole swapping places with the south and vice versa has implications for the Solar System as the wider heliosphere experiences the flip too. The big question is, what is driving the change? Understanding this means understanding why the Sun has this large-scale 'global' change in magnetic activity in the first place.

One of the fundamental rules in the study of electromagnetism is that moving charges create a magnetic field. In 1820, Hans Christian Oersted carried out an experiment with a few simple elements – a battery, a wire and a compass. When the circuit was closed by attaching a battery, the electrons moved along the wire and at the same time the compass flinched to indicate the presence of a magnetic field, only returning to normal when the circuit was broken. On the grand scale of the Sun, the same basic principle applies with no need for a wire or a battery. Because the Sun is mostly made up of moving charged particles its magnetic field is everpresent. Its strength and direction change due to the underlying movements in the Sun. The whole Sun itself is spinning and, without this, the internal dynamo of the Sun would not function and solar activity as we know it would not take place. But where did this key movement start? For that we go back

4.5 billion years to the formation of our whole Solar System.

It turns out that the Sun is still running off the leftover spin from the initial collapse of the cloud of gas and dust that gave birth to the Solar System. Perhaps this was a nearby star exploding, or another large cosmic object moving past it and triggering a collapse. In any case, once some impetus is introduced to this primordial soup of gas and dust leftover from a previous generation star, gravity and angular momentum take hold. The gas cloud collapsed under gravity, spinning faster and faster, getting denser and hotter until the Sun 'switched on' through nuclear fusion as described previously. Without an external stopping force, the Sun will continue its cosmic pirouette in perpetuity.

Since the Sun is a ball of spinning fluid (plasma can be defined as a fluid), it spins faster at the equator than at the poles. There is a ten-day difference, so that a

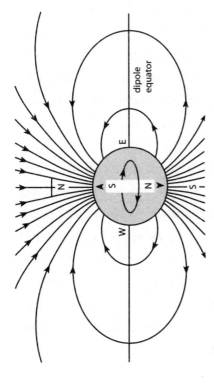

Figure 5: A simplified view of the Earth's magnetic field. If you imagine a bar magnet where the Earth is, picture a north and south pole with field lines connecting them. Any charged particles around the Earth, in the atmosphere and beyond, will follow these field lines to the best of their ability. The entire magnetic system is known as the Earth's magnetosphere and protects us from harmful solar and cosmic radiation. Without this magnetic protection, the solar wind from the Sun would eventually strip away our atmosphere, making the planet inhospitable.

97

sunspot near the poles will take about 35 days to complete one revolution on the photosphere, while a sunspot on the equator will only take 25 days. This differential rotation, as it's known, takes the Sun's standard vertical field lines as seen for a simple bar magnet (see Figure 5) and wraps them up into horizontal magnetic field lines that are stretched and warped (see Figure 6).

Here the plot thickens as parts of this 'wrapped-up' doughnut of magnetic field lines need to 'stand up' to be able to emerge at the Sun's surface as active regions where sunspots are seen. Solar physicists have used very simplified models of moving magnetic fields in plasma to try and understand the inner workings of the Sun, but complex numerical simulations are needed to see what concepts work best. The right one will have to explain all the observed properties of sunspots including their emergence,

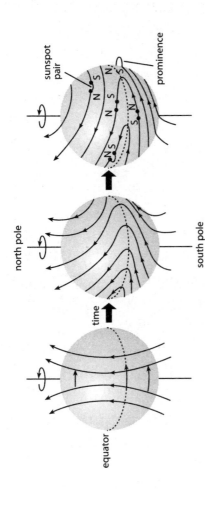

Figure 6: A comic strip version of one of the main processes underpinning the global magnetic system of the Sun. Starting with a normal bar magnet analogy, the lines run vertically from north pole to south. However, the Sun rotates and this winds the magnetic field lines horizontally. From here, on a local level, the magnetic fields try to 'stand up', and do so in sunspot regions which are seen as 'kinks' in the now horizontal field lines. The complete picture of this large-scale cycle of solar magnetic activity remains unconfirmed.

99

relative positions for each sunspot in a pair, disappearance and migration in latitude.

The final part of the picture is the migration of magnetic activity known as **meridional circulation** where sunspots migrate in latitude for the cycle to start again. Thanks to helioseismology solar physicists can extract some information about this from the convection zone. The speeds measured show the sub-surface magnetic activity would take approximately 11 years to go from equator to pole, which matches the solar cycle.

If an explanation for simple vertical magnetic fields changing into areas of emerging magnetic activity just underneath the photosphere is proven, this will bring scientists a great deal closer to understanding the true origins of the Sun's magnetic powers. This understanding may even open up the door to predicting the strengths of impending solar cycles feeding into long term space weather forecasting.

The Sun and our climate

What is the true nature of the Sun's relationship with changes in our climate? The final unanswered question addressed here is complex and has many other questions that can branch off it.

Before digging in, it's important to address our perception of the term 'climate'. Today climate, and more specifically, climate change is a hot topic that governments all over the world are working on to secure the future of life on Earth. The scientific evidence for the short-term rapid, continuing increase in global temperature and subsequent negative effects on our natural environment is clear, and so is the cause. The warming effect of humans burning fossil fuels since industrialisation began in the mid-18th century is 50 times more than the change in the Sun's energy output over the same period, no matter the intensity of solar maxima in the solar cycles occurring during that time.

All that said, unpicking what happens to our climate after the Sun's energy hits Earth's atmosphere is deeply complex. While there is no doubt that we as a species are causing untold damage to our planet through climate change, the Sun is still driving our climate naturally. Astronomers who look for planets orbiting other stars in our galaxy (called exoplanets) sometimes explain that a world might be habitable if it's in the '**Goldilocks Zone**'. This means finding a planet that is not too close to its star and not too far away so that liquid water and therefore life can exist. However, if alien astronomers used the same idea to judge the habitability of Earth from afar, they would end up thinking our world was an ice planet. What the Goldilocks Zone in its most basic form doesn't take into account is an atmosphere and the interactions of a star with that atmosphere. By trapping and re-radiating some of the Sun's energy, a complex climate system is at play. Any alterations in the chemical make-up of our atmosphere can have a profound effect on what way energy is distributed around Earth. Some elements in our atmosphere, such as oxygen and nitrogen, are transparent to infrared energy (heat), while others, such as methane and carbon, are not, trapping the heat and emitting some of it back, heating Earth's surface. The gases that do the trapping are known as the greenhouse gases. Under natural conditions in this stage of Earth's history, there should be a natural equilibrium by having just enough of the Sun's energy radiated back into space.

While the ability to change the amount of some greenhouse gases in our atmosphere lies with us, there is some evidence to show that an indirect effect of the Sun's influence may have some more bearing on one greenhouse gas in particular. During periods of high activity in the Sun, the magnetic bubble it produces called the heliosphere is at its strongest. Normally, cosmic rays from beyond our Solar System strike the Earth regularly, but the increased strength of the heliosphere at solar maxima deflects a great deal of them like a giant shield. Thanks to chemical abundances measured in ice cores and tree rings, scientists have been able to record the variation in the presence of extrasolar cosmic rays striking Earth over thousands of years. They can then compare this to the solar cycle. It has been seen that in most cases, when there is a peak in the solar cycle, the cosmic ray count is drastically reduced.

There have been some localised studies that show that cosmic rays can seed cloud formation on Earth. While sounding incredibly innocuous, these clouds contain water vapour, which is a greenhouse gas. There is no international consensus on the ability for cosmic rays to contribute to cloud formation and even in the small studies available, clouds are not shown to be produced in vast quantities during periods of low solar activity. There is no 'get out of jail free card' for humans' polluting activity, but uncertainty in the chain reaction of solar activity, cosmic rays and cloud formation makes for an intriguing solar conundrum.

The future

Humans have fixated on the Sun through the ages for its life-giving power, its deep astronomical mysteries and its potential to help us fulfil our global energy needs. Looking ahead begs the question: what next for the Sun? While some scientists work on relatively short-term forecasts for space weather and the solar cycle, our Sun has potentially 5 billion years left on the clock for this stage in its evolution.

Astronomers have looked to the vastness of our galaxy to see stars at different stages of their lives, from young to old, star birth to star death, to build 'the big picture'. It's thanks to this expanded view of stars that we understand that after the Sun expends its supply of hydrogen for fusion, it will bloat outwards (sadly engulfing at least Mercury and Venus if not Earth too), and will become a **red giant** star. After this, it will cool and contract to form a **white dwarf** star that will be hot enough for sustained nuclear fusion again for billions more years. By looking at stars of different masses and ages, we are able to predict what the twilight years of our star, the Sun, will be like.

Between short-term solar activity and the Sun's grander life cycle lies an opportunity to look more closely at nearby stars again and compare and contrast local stellar activity. All the tools and techniques developed through the centuries for solar physics can be matched with the increased power of the next generation of mammoth telescopes to look at starspots and stellar cycles of other stars.

Opening that path of discovery would lead to a better understanding of what our Sun could have in store for us in the coming centuries and help humans exist in harmony with it long, long into the future.

Glossary

analemma – the figure-of-eight pattern that results from marking the changing position of the Sun in the sky at the same time and location every day over the course of a year by photography or marking a shadow.

anti-electrons – anti-electrons are the antimatter counterpart of electrons and are often called positrons. They carry a positive charge instead of a negative one.

the Carrington Event – the name given to the white light flare event on the Sun witnessed by Richard Carrington in 1859 as well as referring to the geomagnetic activity experienced on Earth a few days later.

corona – the outer atmosphere of the Sun, derived from the latin for 'crown'. It stretches millions of kilometres outwards from the Sun in all directions and can be seen from Earth during totality in a solar eclipse event.

coronal mass ejection (CME) – huge expulsions of plasma from the Sun that can emanate from active regions where intense magnetic activity is taking place. Depending on the orientation of these regions with Earth, they can reach us and cause geomagnetic activity.

$E=mc^2$ – Einstein's iconic equation stating that matter and energy are interchangeable. According to this equation a small amount of matter is converted into a huge amount of energy during nuclear reactions.

electromagnetic spectrum – the range of wavelengths that make up all electromagnetic emission in the Universe (i.e. all types of light waves). The spectrum spans from long wavelength, low-energy radio waves all the way to short wavelength, high-energy gamma rays.

electron – an elementary subatomic particle in the Standard Model of physics. It carries a negative charge and has a mass that is a very small fraction of the mass of other subatomic particles.

electrostatic force – the repelling or attracting force exerted by two electrically charged objects. The electrostatic force is incredibly strong at the subatomic level requiring a large amount of energy to overcome it to fuse atoms.

Fraunhofer lines – the absorption lines seen in the spectrum of the Sun. Most correspond to elements in the Sun which absorb particular wavelengths of light (colours) leading to these dark lines.

geomagnetic activity – activity related to changes in Earth's own magnetic field usually as a result of activity from the Sun including the solar wind and coronal mass ejections. The activity is broad and includes aurorae and electrical currents sent through the ground.

Goldilocks Zone – the proposed region around a star where a planet could possess liquid water – not too close to the star for the water to boil and not too far away for it to freeze. It is also interchangeable with the term 'habitable zone'.

helioscope – a telescope specially designed for safely projecting a magnified image of the Sun for visual observation.

helioseismology – the study of vibrations or 'ripples' within the Sun that can be measured visually and used to interpret the changing density of the Sun just below the surface.

ionosphere – the term given to the Earth's upper atmosphere between 75 and 1000 kilometres above the Earth's surface. It is susceptible to solar radiation, which can change the density of that region of atmosphere.

magnetic field – a region in which charged objects experience a magnetic force. The strength of the force varies with the strength of the field and can be positive or negative resulting in an attracting or repelling force, respectively. Charged particles such as electrons will follow the direction of a magnetic field where present.

meridional circulation – the cyclical subsurface flow of magnetic activity from the solar equator along the Sun's meridian lines to the poles and back again. It is still unclear exactly how the mechanism works.

NASA's Parker Solar Probe – a spacecraft that launched in 2018 and uses cameras and other instruments to study the Sun up close in flybys. At its closest, it will be 6 million kilometres from the Sun – in comparison, the innermost planet Mercury is on average 57.9 million kilometres distant.

neutrinos – 'ghost-like' particles that pass through matter with little to no interactions. They are nearly massless and carry no charge, making them extremely difficult to detect.

nuclear fusion – a type of nuclear reaction that takes place when the nuclei of two atoms are fused together and transform a small amount of their mass into energy and a heavier element. It is through this process of nuclear fusion of the fundamental elements of the Universe (hydrogen and helium) that all the elements we see around us were formed in the heart of stars.

nuclei – the centres of all atoms in the Universe. They consist of particles called protons and neutrons. The number of protons in a single nucleus defines which element an atom is.

parallax – difference in the apparent position of an object viewed along two different lines of sight with respect to more distant background objects. It can be used to measure distances to far away objects in space.

photoheliograph – an instrument that was used for the capture of photographs of the full disc of the Sun in white light from the 19th century onwards. The basic original design included a combination of a telescope, shutter mechanism and a photographic plate.

photon – packets of energy that make up all forms of radiation across the electromagnetic spectrum, including the light the human eye can see.

plasma – sometimes referred to as a fourth state of matter (after solid, liquid and gas), in terms of solar and stellar physics, plasma is what happens when gases are energised to such a degree that electrons are stripped from atoms leaving a mix of nuclei and free electrons.

prominence – a loop of plasma from the Sun that emerges from the photosphere and stretches up into the Sun's chromosphere, on into the corona and back down again. Prominences are only seen from Earth at the edge of the Sun due to the contrast against empty space.

red giant – a stage of life a star reaches when it has used up the hydrogen in its core through nuclear fusion. The star bloats outward into the red giant phase and starts fusing hydrogen in the region around its helium core.

solar activity – an umbrella term for magnetic activity that grows and recedes within the Sun over the course of the 11-year solar cycle. The activity primarily manifests as active regions on the Sun's surface containing groups of sunspots, which in turn can lead to flares and CMEs.

solar wind – a continuous stream of charged particles that leave the Sun via open magnetic fields. The solar wind speed varies depending on magnetic activity on the Sun and extends all the way out to the outer regions of the Solar System.

spectrograph – an instrument that splits incoming light into its component wavelengths. This instrument, an evolution of the spectroscope, also includes a camera to capture the resulting light spectrum so that it can be analysed after the initial observation.

spectrohelioscope – this instrument allows the user to scan the surface of the Sun in particular wavelengths to see just one particular layer of the Sun in more specific detail. It was later adapted with camera technology to become a spectroheliograph.

spectroscope – an instrument that disperses light into its various wavelengths, usually using a prism, a grating or a combination of the two.

sunspot – a darker, cooler area on the surface of the Sun. These features are transient and vary in size, position and longevity. They generally appear in pairs or groups.

totality – the period of time that the Sun is completely obscured by the Moon during a total solar eclipse. This period can last up to seven and a half minutes.

white dwarf – for stars with a mass in or around the mass of the Sun, the red giant phase is followed by the evolution to a white dwarf star. The star expels its outer layers and contracts to form a relatively small and dense white star that will radiate for billions of years until all that is left is a cold core.

Royal Observatory
Greenwich Illuminates

Stars
by Dr Greg Brown
978-1-906367-81-7

Planets
by Dr Emily Drabek-Maunder
978-1-906367-82-4

Black Holes
by Dr Ed Bloomer
978-1-906367-85-5

Space Exploration
by Dhara Patel
978-1-906367-88-6